P9-CNA-797

STUDENT GUIDE

WHAT DOES THE DATA SAY?

GRAPHS AND AVERAGES

WITHDRAWN

MathScape™
SEEING AND THINKING MATHEMATICALLY

PHASE**ONE**
Measures of Central Tendency

Super Data Company collects data, shows data on graphs, and analyzes data. As an apprentice statistician, you will conduct a survey, create a graph of the data you collect, and analyze it. These are important skills for statisticians to know.

How can we use data to answer questions about the world around us?

WHAT DOES THE DATA SAY?

PHASE**TWO**

Representing and Analyzing Data

You will explore bar graphs. You will create single bar graphs and identify errors in several bar graphs. You will also investigate how the scale on a bar graph can affect how the data is interpreted. At the end of the phase, you will conduct a survey of two different age groups. Then you will make a double bar graph to represent the data and compare the opinions of the two groups.

PHASE**THREE**

Progress over Time

Start off with a Memory Game in which you find out if your memory improves over time and with practice. You will create a broken-line graph to show your progress. Then you will analyze the graph to see if you improved, got worse, or stayed the same. The phase ends with a project in which you measure your progress at a skill of your choice.

PHASE**FOUR**

Probability and Sampling

You will investigate the chances of choosing a green cube out of a bag of green and yellow cubes. Then you will explore how changing the number of cubes in the bag can change the probability of picking a green cube. At the end of the phase, you will apply what you've learned so far by helping one of Super Data Company's clients solve the Jelly Bean Bag Mix-Up.

PHASE ONE

To: Apprentice Statisticians
From: President, Super Data Company

Welcome to Super Data Company! In your new job as an apprentice statistician, you will conduct and analyze surveys for our many customers.

There are different kinds of questions that you can ask in a survey. There is also a lot of information that you can get from the results of surveys. In your first assignment, you will conduct a survey and collect data about your classmates.

A statistician collects and organizes data. Surveys, questionnaires, polls, and graphs are tools that statisticians use to gather and analyze the information.

In Phase One, you will begin your new job as an apprentice statistician by collecting data about your class. You will learn ways to organize the data you collect. Then you will analyze the data and present your findings to the class.

Measures of Central Tendency

WHAT'S THE MATH?

Investigations in this section focus on:

COLLECTING DATA

- Conducting surveys to collect data
- Collecting numerical data

GRAPHING

- Making and interpreting frequency graphs

ANALYZING DATA

- Finding the mean, median, mode, and range of a data set
- Using mean, median, mode, and range to analyze data

mathscape1.com/self_check_quiz

1 Class Survey

ANALYZING DATA TO FIND MODES AND RANGES

How well do you know your class? Taking a survey is one way to get information about a group of people. You and your classmates will answer some survey questions. Then you will graph the class data and analyze it. You may be surprised by what you find out about your class.

Find the Mode and Range

How can you find the mode and range for a set of data?

The data your class tallied from the Class Survey Questions is a list of numbers. The number that shows up most often in a set of data like this is called the *mode.* The *range* is the difference between the greatest number and the least number in a set of data. For the data in the class frequency graph you see here, the mode is 10. The range is 9.

Look at the frequency graph your class created for Question A. Find the mode and the range for your class data.

How Many Glasses of Soda We Drink

```
                                   X
                                   X
   X                               X
   X                               X
   X                   X           X
   X                   X           X
   X       X           X           X
   X       X           X           X
   X       X           X           X
X  X       X           X           X
-----------------------------------
1  2   3   4   5   6   7   8   9   10
          Number of glasses
```

Analyze the Class Data

How can you use mode and range to analyze data?

Your teacher will give your group the class's responses for one of the survey questions you answered at the beginning of the lesson. Follow these steps to find out everything you can about your class.

1 Create a frequency graph of the data.

a. Include everyone's answer on your graph.

b. Don't forget to label the graph and give it a title.

2 Analyze the data from your graph.

a. Find the mode.

b. Find the range.

Write About the Class Data

Write a summary that clearly states what you learned about your class from the data. Be sure to include answers to the following questions:

- What does the data tell you about the class? Make a list of statements about the data. For example, "Only one student in the class has 7 pets."
- What information did you find out about the class from the mode and range?

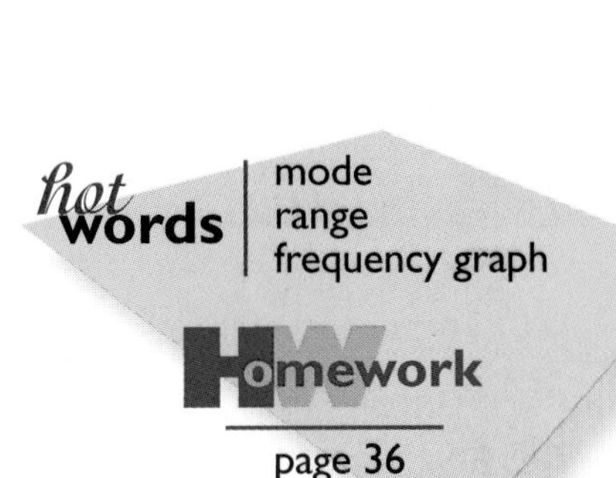

2 Name Exchange

EXPLORING MEANS AND MEDIANS

One of the questions often asked about a set of data is, "What is typical?" You have learned to find the mode of a list of numbers. Two other measures of what's typical are the *mean* and the *median.* Here you will use mean, median, and mode to analyze data on the names in your class.

Find the Mean

How can you find the mean length of a name in the class?

One way to find the mean length of a set of first names is to do the Name Exchange. Follow these steps to find the mean, or average, length of the first names of members in your group.

1. Write each letter of your first name on a different sheet of paper.
2. Members of your group should exchange just enough letters so that either:

   a. each member has the same number of letters, or

   b. some group members have just *one* letter more than other members.

   You may find that some members of the group do not need to change letters at all.
3. Record the mean, or average, length of the first names in your group.

There are 5 girls in the group and these are their names.

Sherry	Lorena	Natasha	Daniella	Ann
6	6	7	8	3

Natasha and Daniella gave letters to Ann, so everyone would have 6 letters.

Sherry	Lorena	Natash-a	Daniel-la	Ann+a+la
6	6	6	6	

The mean length for the group is 6.

How does the mean for your group compare to the mean for the class?

Find the Median

How can you find the median for a data set?

When the numbers in a set of data are arranged in order from least to greatest, the number in the middle is the median. If there is an even number of numbers in a set of data, the median is the mean of the two middle numbers. Use the frequency graph your class made for the lengths of names to answer these questions.

- What is the median length of a first name for your class?
- What does the median tell you?

How does the mean compare to the median for the class?

Write About the Class Data

You have learned about the mean, median, mode, and range. Think about what you have learned to answer the following questions about your class:

- What do each of the measures (mean, median, mode, and range) tell you about the lengths of the first names in the class?
- Which of the measures (mean, median, or mode) do you think gives the best sense of what is typical for the class? Why?
- What are some situations where it would be helpful to know the mean, median, mode, or range for the class?

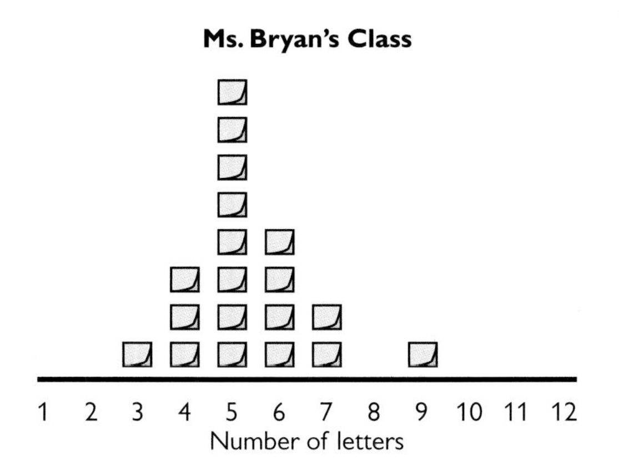

3 TV Shows

INVESTIGATING RATINGS AND DISTRIBUTIONS

Rating scales are often used to find out about people's opinions. After your class rates some television shows, you will look at some data on how another group of students rated other television shows. Then you will apply everything you have learned so far to conduct and analyze a survey of your own.

What information can you get by analyzing the distribution of data in a graph?

Analyze Mystery Graphs

The graphs below show how some middle school students rated four TV shows. Use the information in the graphs to answer the following questions:

- Overall, how do students feel about each show?
- Do the students agree on their feelings about each show? Explain your answer.
- Which TV shows that *you* watch might give the same results if your classmates rated the shows?

Mystery Graphs

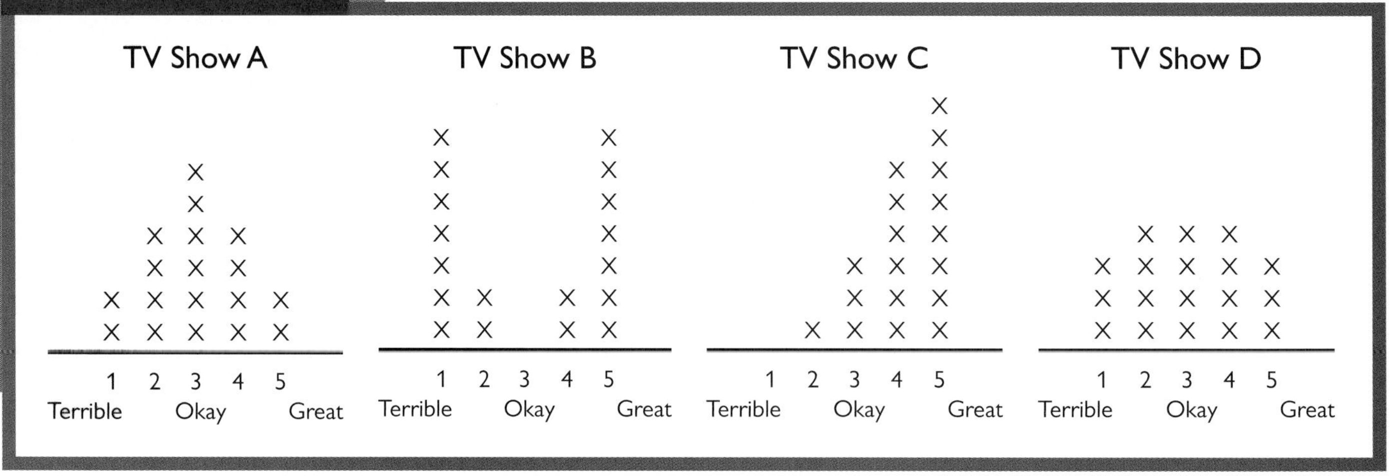

TV Show A

1 2 3 4 5
Terrible Okay Great

TV Show B

1 2 3 4 5
Terrible Okay Great

TV Show C

1 2 3 4 5
Terrible Okay Great

TV Show D

1 2 3 4 5
Terrible Okay Great

Collect and Analyze Data, Part 1

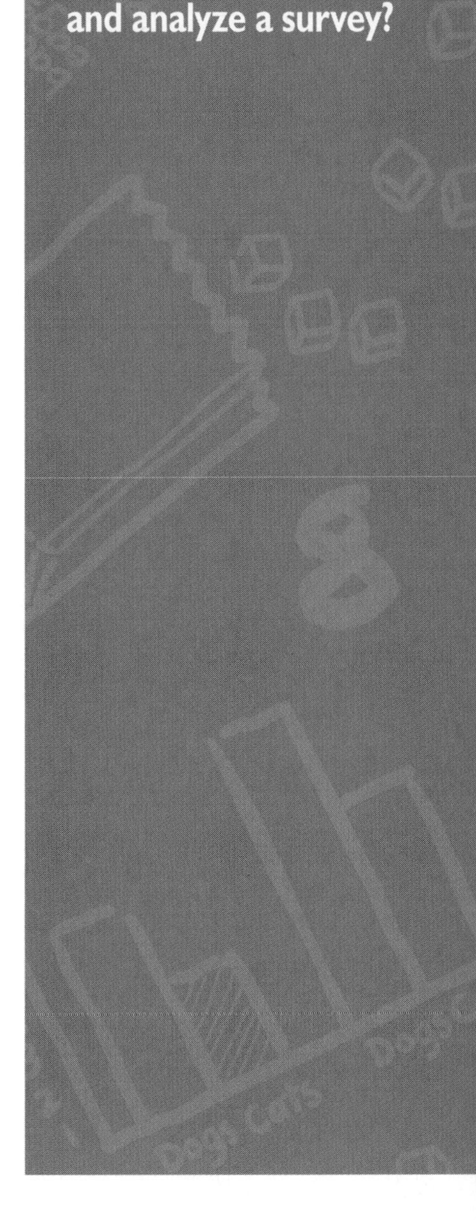

How can you conduct and analyze a survey?

Now it's your turn! You will apply what you have learned about collecting, representing, and analyzing data to find out about a topic of your choice. Follow these steps.

1 Make a data collection plan

Choose a topic	On what topic would you like to collect data?
Choose a population to survey	Whom do you want to survey? For example, do you want to ask 6th graders or 1st graders? How will you find at least 10 people from your population to survey?
Write survey questions	Write four different survey questions that can be answered with numbers. At least one of the questions should use a rating scale. Make sure that the questions are easy to understand.
Identify an audience	Who might be interested in the information you will collect? Why might they be interested?

2 Collect and represent data

Collect data	Collect data for just one of the survey questions. Ask at least 10 people from the population you chose. Record your data.
Graph data	Create an accurate frequency graph of your data.

3 Analyze the data

Write a report that answers these questions:	What are the mean, median, mode, and range? How would you describe the distribution, or shape, of the data? What did you find out? Make a list of statements that are clearly supported by the data.

page 38

PHASE TWO

To: Apprentice Statisticians
From: President, Super Data Company

I hope you are enjoying your work with Super Data Company!

Super Data Company is working on a new exhibit for the local zoo. The exhibit will make it easy for visitors to find out and compare interesting facts about animals. A bar graph is a useful way to represent data. In your next assignment, you will make bar graphs about zoo animals.

In this phase, you will investigate and create bar graphs. You will also learn how the scales on a bar graph can affect how the information is interpreted.

Graphs are used to represent information about many things, such as advertisements, test results, and political polls. Where have you seen bar graphs? What type of information can be shown on a bar graph?

Representing and Analyzing Data

WHAT'S THE MATH?

Investigations in this section focus on:

COLLECTING DATA

- Conducting surveys to collect data
- Collecting numerical data

GRAPHING

- Making and interpreting single and double bar graphs

ANALYZING DATA

- Comparing data to make recommendations

4 Animal Comparisons

INVESTIGATING BAR GRAPHS AND SCALES

Graphs are used in many different ways, like showing average rainfall or describing test results. Here you will explore how the scale you choose changes the way a graph looks as well as how it affects the way people interpret the data.

How can you choose scales to accurately represent different data sets in bar graphs?

Represent Data with Bar Graphs

1. Choose two sets of data about zoo animals to work with.
2. Make a bar graph for each set of data you chose. Follow the guidelines below when making your graphs:
   - **a.** All the data must be accurately represented.
   - **b.** Each bar graph must fit on an $8\frac{1}{2}$" by 11" sheet of paper. Make graphs large enough to fill up at least half the paper.
   - **c.** Each bar graph must be labeled and easy to understand.
3. After you finish making the bar graphs, describe how you chose the scales.

What is important to remember about showing data on a bar graph?

Animal	Weight (pounds)
Sea cow	1,300
Saltwater crocodile	1,100
Horse	950
Moose	800
Polar bear	715
Gorilla	450
Chimpanzee	150

Animal	Weight (ounces)
Giant bat	1.90
Weasel	2.38
Shrew	3.00
Mole	3.25
Hamster	4.20
Gerbil	4.41

Animal	Typical Number of Offspring (born at one time)
Ostrich	15
Mouse	30
Python	29
Pig	30
Crocodile	60
Turtle	104

Animal	Speed (miles per hour)
Centipede	1.12
House spider	1.17
Shrew	2.5
House rat	6.0
Pig	11.0
Squirrel	12.0

Investigate Scales

How does changing the scale of a graph affect how people interpret the data?

The scale you choose can change the way a graph looks. To show how this works, make three different graphs for the data in the table.

Number of Visitors per Day at the Zoo

Zoo	Visitors per Day
Animal Arc Zoo	1,240
Wild Animal Park	889
Zooatarium	1,573

1 Make one graph using a scale that gives the most accurate and fair picture of the data. Label this Graph A.

2 Make one graph using a scale that makes the differences in the numbers of visitors look smaller. Label this Graph B.

3 Make one graph using a scale that makes the differences in the numbers of visitors look greater. Label this Graph C.

Write About Scales

Look at the three graphs you made for Number of Visitors per Day at the Zoo.

- Explain how you changed the scale in Graphs B and C.
- Describe some situations where someone might want the differences in data to stand out.
- Describe some situations where someone might want the differences in data to be less noticeable.
- List tips that you would give for choosing accurate scales when making a graph.
- Explain how you check a bar graph to make sure it accurately represents data.

hot words | bar graph

Homework page 39

5 Double Data

CREATING AND INTERPRETING DOUBLE BAR GRAPHS

A double bar graph makes it easy to compare two sets of data. After analyzing a double bar graph, you will make recommendations based on the data shown. Then you will be ready to create and analyze your own double bar graph.

How can you use a double bar graph to compare two sets of data?

Analyze the Double Bar Graph

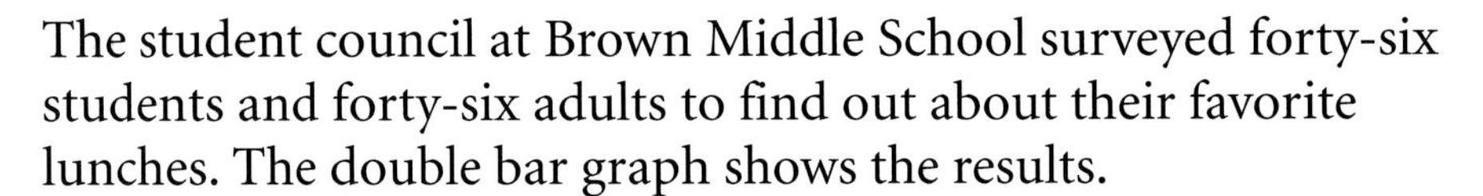

The student council at Brown Middle School surveyed forty-six students and forty-six adults to find out about their favorite lunches. The double bar graph shows the results.

- What is the most popular lunch for students? for adults?
- What is the least popular lunch for students? for adults?
- Why is there no bar for adults where hamburgers are shown?

Make Recommendations

Use the data in the double bar graph to make recommendations to the student council about what to serve for a parent-student luncheon. Be sure to use fractions to describe the data.

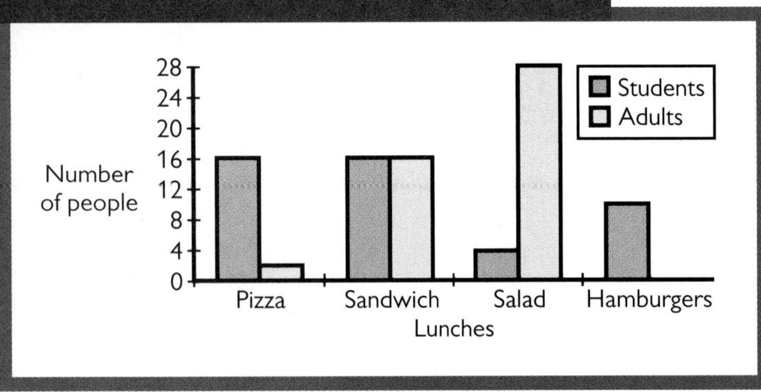

- What would you recommend that the student council serve at the parent-student luncheon?
- What other things should the student council think about when choosing food for the luncheon?

Make Double Bar Graphs

How can you create a double bar graph to compare two sets of data?

The table shown gives data on the number of hours people in different age groups sleep on a typical night.

Hours of Sleep	6-Year-Olds	12-Year-Olds	14-Year-Olds	Adults
5	0	0	3	9
6	0	2	7	15
7	0	16	24	20
8	0	19	36	20
9	3	25	5	9
10	4	12	5	3
11	31	4	0	4
12	37	2	0	0
13	5	0	0	0

1. Choose two columns from the table.
2. Make a double bar graph to compare the data. Be sure your graph is accurate and easy to read.
3. When you finish, write a summary of what you found out.

Write About Double Bar Graphs

Students in Ms. Taylor's class came up with this list of topics.

Answer these questions for each topic:

- Could you make a double bar graph to represent the data?
- If it is possible to make a double bar graph, how would you label the axes of the graph?

A. Number of students and teachers at our school this year and last year.

B. Heights of students in the same grade.

C. Time students spend doing homework and time they spend playing sports.

D. Number of hours students watch television and number of televisions in their homes.

E. Number of miles students travel to school.

6 Across the Ages

MAKING COMPARISONS AND RECOMMENDATIONS

Do you think middle school students and adults feel the same about videos? In this lesson, you will analyze the results of a survey about videos. Then you will conduct your own survey to compare the opinions of two age groups.

How can you use double bar graphs to compare the opinions of two age groups?

Compare Opinion Data from Two Age Groups

The parent-student luncheon was a huge success. The student council has decided to hold a parent-student video evening. They surveyed about 100 middle school students and about 100 adults to find out their opinions of four videos. The double bar graphs on the handout Video Rating Scale show how students and adults rated each video. Use the graphs to answer these questions:

- How do middle school students feel about each video? Adults?
- Which video do students and adults disagree about the most? Explain your thinking.
- Which video do students and adults agree about the most? Explain your thinking.

Make Recommendations

Make recommendations to the Student Council at Brown Middle School. Be sure the data supports your recommendations.

- Which video should the council choose for a students-only video evening? Why?
- Which video should the council choose for an adults-only video evening? Why?
- Which video should the council choose to show at a parent-student video evening? Why?

Collect and Analyze Data, Part 2

How can you compare data from two different groups?

In Lesson 3, you conducted your own survey. This time you will conduct the survey again, but you will collect data from a different age group. Then you will compare the results of the two surveys. You will need your survey from Lesson 3 to complete this activity.

1 Choose a new age group.

a. What age group do you want to survey?

b. How will you find people from the new age group to survey? In order to compare the two groups fairly, you will need to use the same number of people as you did for the first survey.

2 Make a prediction about the results.

a. How do you think people in the new group are likely to respond to the survey?

b. How similar or different do you think the responses from the two groups will be?

3 Collect and represent the data.

a. Collect data from the new group. Ask the same question you asked in Lesson 3. Record your data.

b. Create a single bar graph to represent the data from the new group.

c. Create a double bar graph to compare the data from the two groups.

d. Explain how you chose the scales for the two graphs.

4 Analyze the data.

Write a report that includes the following information:

a. What are the mean, median, mode, and range for the new age group? What do these measures tell you?

b. How do the responses for the two groups compare? Make a list of comparison statements that are clearly supported by the data.

c. How do the results compare with your predictions?

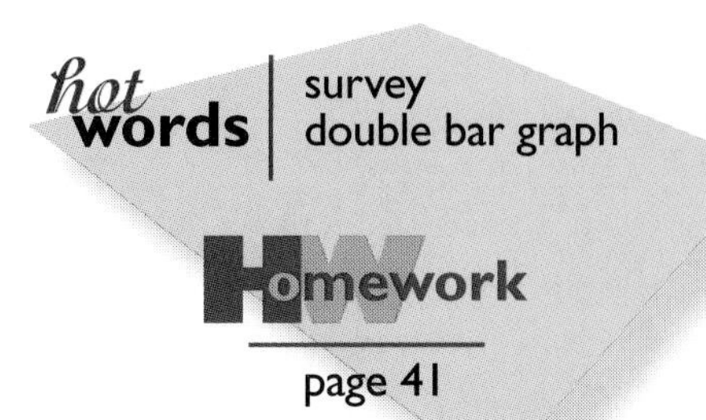

PHASE THREE

To: Apprentice Statisticians
From: President, Super Data Company

You have been doing a great job conducting surveys and making graphs. Next, you are going to analyze progress over time.

When people learn new skills, like typing or playing a musical instrument, they need to practice a lot. It's hard work, so they want to know whether the practicing is paying off. Statistics can help measure progress. In your next assignment, you will collect data and analyze your performance at several skills.

Have you ever tried to learn a new skill, such as typing, juggling, or making free throws? How can you tell if you are improving? How can you tell if you are getting worse?

In Phase Three, you will measure your progress at some skills. Then you will use statistics to help you see whether you have been improving, staying the same, going up and down, or getting worse.

Progress over Time

WHAT'S THE MATH?

Investigations in this section focus on:

DATA COLLECTION

- Collecting numerical data

GRAPHING

- Making and interpreting broken-line graphs
- Using broken-line graphs to make predictions

DATA ANALYSIS

- Finding mean, median, mode, and range in a data set
- Using mean, median, mode, and range to analyze data

Are You Improving?

USING STATISTICS TO MEASURE PROGRESS

Learning new skills takes a lot of practice. Sometimes it's easy to tell when you are improving, but sometimes it's not. Statistics can help you measure your progress. To see how this works, you are going to practice a skill and analyze how you do.

Graph and Analyze the Data

How can a broken-line graph help you see whether you improved from one game to the next?

After you play the Memory Game with your class, make a broken-line graph to represent your data.

1. For each game, or trial, plot a point to show how many objects you remembered correctly. Connect the points with a broken line.
2. When your graph is complete, analyze the data. Find the mean, median, mode, and range.
3. Write a summary of your findings. What information do the mean, median, mode, and range tell you about your progress? Overall, do you think you improved? Use data to support your conclusions. How do you predict you would do on the 6th game? the 10th one? Why?

The Memory Game

How to play:

1. You will be given 10 seconds to look at pictures of 9 objects.
2. After the 10 seconds are up, write down the names of the objects you remember.
3. When you look at the pictures again, record the number you remembered correctly.

Sample Broken-Line Graph for the Memory Game

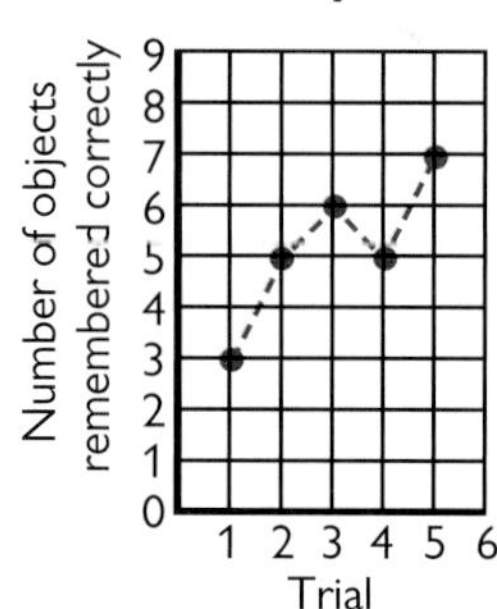

Compare Data

How can you compare two sets of data?

Two students played 5 trials of the Memory Game. In each trial, they looked at pictures of 9 objects for 10 seconds. The table shows their results. Use the table to compare Tomiko's and Bianca's progress.

1 Make broken-line graphs to show how each student did when playing the Memory Game.

2 Find the mean, median, mode, and range for each student.

3 Analyze the data by answering the following questions:

a. Which student do you think has improved the most? Use the data to support your conclusions.

b. How many objects do you think each student will remember correctly on the 6th game? the 10th game? Why?

Tomiko's and Bianca's Results

Number of Objects Remembered Correctly

Trial Number	Tomiko	Bianca
1	3	4
2	3	6
3	4	5
4	6	9
5	8	7

Design an Improvement Project

Now it's time to apply what you have learned. Think of a skill you would like to improve, such as juggling, running, balancing, or typing. The Improvement Project handout will help you get started.

1 Write a plan for a project in which you will practice the skill (see Step 1 on the handout).

2 Over the next 5 days, practice the skill and record your progress (see Step 2 on the handout).

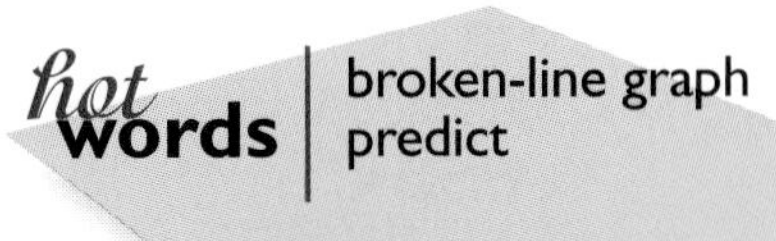

page 42

8 How Close Can You Get?

GRAPHING AND ANALYZING ERRORS

In the last lesson, you kept track of your progress in the Memory Game. Now you will play an estimation game in which you try to get closer to the target with each turn. You will keep track of your progress and graph your errors to see if you improved.

How can you represent your errors so you can easily see the progress you have made?

Graph and Analyze Progress

After you play the Open Book Game, make a broken-line graph to represent your data. Be sure to label both axes.

1. For each trial, plot a point to represent the error. Connect the points with a broken line.
2. When your graph is complete, analyze the data. Find the mean, median, mode, and range.
3. Write a summary of your progress. What information do the mean, median, mode, and range tell you about your progress? Overall, do you think you improved? Make sure to use data to support your conclusions. How do you predict you would do on the 6th game? the 10th one? Why? How would you describe your progress from game to game?

The Open Book Game

How to play:

1. Your partner will tell you a page to turn to in the book. Try to open the book to that page without looking at the page numbers.
2. Record the page number you tried to get (Target) and the page you opened the book to (Estimate).
3. Figure out and record how close you were to the target page (Error) by finding the difference between the Target and the Estimate. Subtract the lesser number from the greater number.
4. For each trial, your partner will tell you a different page number. After 5 trials, switch roles.

Investigate Mixed-up Data

Kim loves to run and wants to get faster. Every day for 20 days she ran around her block and timed how long it took. She kept track of her progress on a broken-line graph and wrote about it in her journal. Unfortunately, her journal fell apart, and all the entries are out of order. Can you figure out which journal entry goes with which days?

How can you use what you have learned about graphing to sort out some mixed-up information?

Kim's Journal Entries

A Practicing is paying off. I'm making steady progress.

B I'm disappointed because I'm not making progress. At least I'm not getting worse.

C Wow. I've made my biggest improvement yet.

D I've been doing worse. I hope it's because I have a bad cold.

E I don't know what's going on. The time it takes me to run has been going up and down from one day to the next.

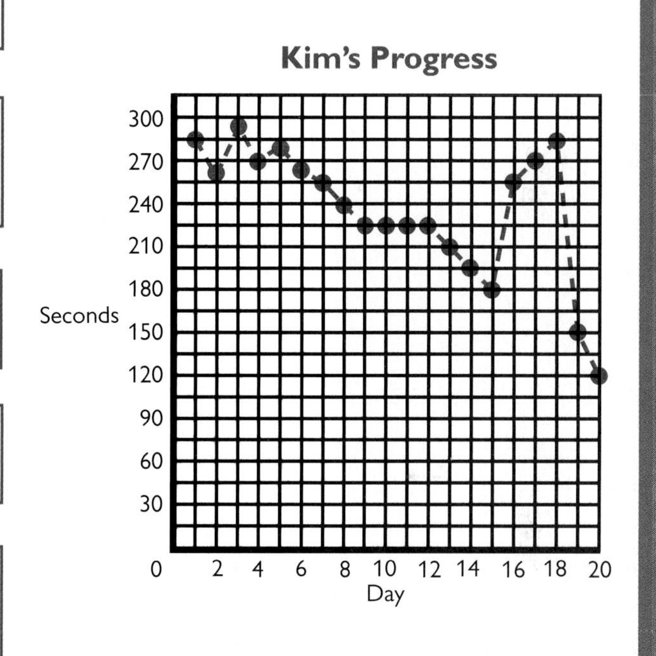

Read over the entries and examine the graph to help you answer these questions:

- Which days do you think each journal entry describes? Why? Tip: Each entry describes Kim's progress over 2 or more days.
- How did you figure out which entry went with which days on the graph?
- How would you describe Kim's overall progress for 20 days?
- How do you think Kim would do on days 21, 22, 23, and 24? Why? If you were Kim, what would you write in your journal about your progress on those days?

Stories and Graphs

INTERPRETING MULTIPLE REPRESENTATIONS OF DATA

Can you look at a graph and figure out what story it tells? In this lesson, you will interpret unlabeled graphs to figure out which ones match different people's descriptions of learning skills. Then you will compare and analyze graphs of progress and predict future performance.

Match Descriptions to Graphs

How can you figure out what stories a graph might represent?

Six students worked on improving their skills. They measured their progress by timing themselves. Then they wrote descriptions of their progress. They also graphed their data, but forgot to put titles on their graphs. Read the descriptions and study the graphs on the handout of Graphs of Students' Progress.

- Figure out which graph goes with which student. Explain your reasoning.
- Write a title for each graph.
- The extra graph belongs to Caitlin. Choose a skill for Caitlin. Then write a description of her progress.

Descriptions of Students' Progress

I've been practicing running fast around my block. I keep getting faster and faster. Jian

I've been practicing balancing on one foot with my eyes closed. I did better for a few days, then I did worse, then I did better again. Natasha

I'm working on rollerblading fast around my block. I'm not improving at all. I had my best time on my first day and the worst time on my last day. Miguel

Oops. I forgot to write a description. Caitlin

I'm practicing balancing on one foot with my eyes closed. I've been making steady progress. Keishia

I'm practicing juggling as long as I can without dropping a ball. I did great the first day, then I got worse. Then I stayed the same. At the end I was doing better. Terrence

Analyze Data from the Improvement Project

After practicing a skill for 5 days, how can you tell if you improved?

Now it's time to look at the data you've been collecting for the last 5 days. You will need the Improvement Project handout. Enter your data for each day in the table on your handout. Then analyze the data.

- Complete the table by entering the range, mean, mode, and median for each day's data.
- Look at your completed table. Use data you choose from the table to make two graphs. For example, you may want to graph the mean, or the highest score, or the total score for each day.

Write About the Improvement Project

Write a report to share your project with the class. Include answers to the following questions in your report:

- How did you decide which data to graph? Does one graph show more improvement than the other?
- Describe your progress from day to day.
- Overall, do you think you improved? Use the data to support your conclusions.
- How do you think you will do on the 6th day? Why?
- What mathematics did you use in the project?

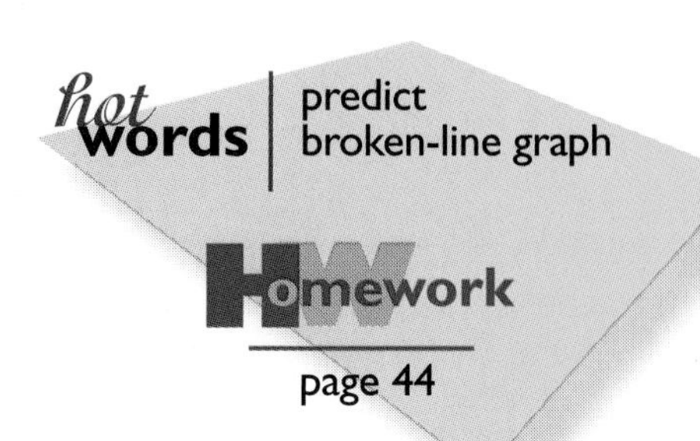

PHASE FOUR

To: Apprentice Statisticians
From: President, Super Data Company

I hope you have enjoyed your work at Super Data Company so far!

People often talk about the likelihood of things happening (there's an 80% chance of rain; the Lions are favored to win). Being able to predict the probability that a particular event will occur can be very useful. In your next assignment, you will find the probability of picking a green cube out of a bag of colored cubes. You will need to use your data-collection and analysis skills in the investigation.

Have you ever decided to wear a raincoat because the weather report said it was likely to rain? Have you ever bought a raffle ticket because you thought your chances of winning were good? If you have, then you were basing your decisions on the probability that a specific event would occur.

Probability is the mathematics of chance. In this phase, you will investigate probability by playing some games of chance.

Probability and Sampling

WHAT'S THE MATH?

Investigations in this section focus on:

DATA COLLECTION

- Collecting and recording data
- Sampling a population

DATA ANALYSIS

- Using the results of sampling to make a hypothesis
- Using bar graphs to make an informed prediction

DETERMINING PROBABILITY

- Describing probabilities
- Calculating theoretical and experimental probabilities

10 What Are the Chances?

DETERMINING PROBABILITY

Probability is the mathematics of chance. Raffles involve chance. The tickets are mixed together, and one ticket is picked to be the winner. Here you will play a game that is similar to a raffle. Then you will figure out the probability of winning.

Analyze Data from a Game of Chance

How can you use data to predict the probability of picking a green cube?

The Lucky Green Game is a game of chance. The chances of winning could be very high or very low. You will conduct an investigation to find out just how good the chances of winning really are.

1. Play the game with your group. Be sure each player takes 5 turns. Record each player's results in a table.
2. After each player has taken 5 turns, answer the following questions:
   - **a.** How many greens do you think are in the bag? Use your group's results to make a hypothesis. Be sure to explain your thinking.
   - **b.** Which of these words would you use to describe the probability of picking a green cube?

   Never • Very Unlikely • Unlikely • Likely • Very Likely • Always

How can you analyze the class results?

The Lucky Green Game

Your group will be given a bag with 5 cubes in it. Do not look in the bag! Each player should do the following steps 5 times.

1. Pick one cube from the bag without looking.
2. If you get a green, you win. If you get a yellow, you lose. Record your results.
3. Put the cube back and shake the bag.

Analyze Data for a Different Bag of Cubes

What conclusions can you draw from another class's data?

Ms. Ruiz's class did an experiment with a bag that had 100 cubes in two different colors. Each group of students took out 1 cube at a time and recorded the color. Then they put the cube back in the bag. Each group did this 10 times. The groups put their data together in a class table shown on the handout, A Different Bag of Cubes. Help Ms. Ruiz's class analyze the data by answering these questions:

1 What are the mode, mean, median, and range for the numbers of cubes of each color that were drawn?

2 Based on the whole class's data, what is the experimental probability of picking each color?

3 Here is a list of bags that the class might have used in the experiment. Which bag or bags do you think the class used? Explain your thinking.

a. 50 red, 50 blue
b. 20 red, 80 blue
c. 80 red, 20 blue
d. 24 red, 76 blue
e. 70 red, 30 blue
f. 18 red, 82 blue

Types of Probabilities

Experimental probabilities describe how likely it is that something will occur. Experimental probabilities are based on data collected by conducting experiments, playing games, and researching statistics in books, newspapers, and magazines.

The experimental probability of getting a cube of a particular color can be found by using this formula:

$$\frac{\text{Number of times a cube of a particular color was picked}}{\text{Total number of times a cube was picked}}$$

Theoretical probabilities are found by analyzing a situation, such as looking at the contents of the bag.

The theoretical probability of getting a cube of a particular color can be found by using this formula:

$$\frac{\text{Number of cubes of that color in the bag}}{\text{Total number of cubes in the bag}}$$

experimental probability
theoretical probability

page 45

11 Changing the Chances

EXPERIMENTING WITH PROBABILITY

Does having more cubes in the bag improve your chances of winning? In this lesson, you will change the number of green and yellow cubes. Then you will play the Lucky Green Game to see if the probability of winning has gotten better or worse.

How does changing the number of cubes in the bag change the probability of winning?

Compare Two Bags of Cubes

In Lesson 10, you found the probability of winning the Lucky Green Game. Now, you'll conduct an experiment to see how changing the number of cubes in the bag changes the chances of winning. Follow these steps to find out:

1. Change the number of cubes in the bag you used in Lesson 10 (Bag A), so that it contains 6 green cubes and 4 yellow cubes. Call this new bag, Bag B.
2. Make a hypothesis about which bag (Bag A or Bag B) gives you a better chance of picking a green cube. Explain your reasoning.
3. Collect data by playing the Lucky Green Game (see page 30). Make sure each player takes 5 turns! Record your results on the handout Changing the Cubes Recording Sheet.

Summarize the Data

After your group finishes the experiment, write a summary of your data that includes the following information:

- What were the range, mode, and mean number of greens?
- What was the experimental probability of picking a green cube?
- Did your results support your hypothesis about which bag (A or B) gives you a better chance of picking a green cube? Why or why not?

Rank the Bags

How can you compare bags with different numbers of cubes?

Mr. Chin's class wants to investigate the chances of winning with different bags of cubes. The table below shows the numbers of cubes in the bags Mr. Chin's class plans to use.

More Bags of Cubes

Bag	Green Cubes	Yellow Cubes	Total Number of Cubes
B	6	4	10
C	7	13	20
D	14	6	20
E	13	27	40
F	10	30	40

1. Choose one of the bags (except Bag B). If you picked a cube from that bag 100 times, how many times do you think you would get a green cube? Why?
2. For each bag, find the theoretical probability of picking a green cube. Explain how you figured it out.
3. Rank the bags from the best chance of getting a green cube to the worst chance of getting a green. (Best = 1, Worst = 5) Be sure to explain your answer.
4. After you finish ranking the bags, make a new bag of cubes that will give you a better chance of getting a green than the second-best bag, but not as good as the best bag. How many green and yellow cubes are in the new bag? Explain.

Make Generalizations

Use the results of your data to answer these questions:

- A class did an experiment with one of the bags shown in the table. In 100 turns, they got 32 yellow cubes. Which bag or bags do you think it is most likely that they used? Why?
- What generalizations would you make about how to determine which bag of cubes gives you a better chance of picking a green cube?

12 Which Bag Is Which?

APPLYING PROBABILITY AND STATISTICS

In the last two lessons, you used a method called *sampling* when you made predictions. Here you will use sampling again to predict what's in the bag, but this time you will need to share your findings with the rest of the class in order to be sure.

Investigate the Jelly Bean Bag Mix-Up

How can you use what you have learned about sampling to make predictions?

The graphs on the handout Jelly Bean Bag Combinations show how many jelly beans are in each bag. Each group in your class will get one of the bags to sample. Can you tell which graph matches your bag?

1 Collect data by sampling your bag.

2 Compare your data to the bar graphs on Jelly Bean Bag Combinations. Which bag do you think you have? Write down why you think your group has that bag. If you are not sure, explain why.

Sampling the Jelly Bean Bags

How to sample:

Each student should do the following steps 6 times (that is, take 6 samples):

1. Pick one cube from the bag without looking.
2. Record which color you got in a table like the one shown.
3. Put the cube back and shake the bag before taking the next sample.

Student	Cherry (Red)	Blueberry (Blue)	Lemon (Yellow)	Lime (Green)
Marie Elena	I	I I I	I	I
Ricardo	I I	I I I	I	
Myra	I	I I	I I	I
Ursula	I	I I	I	I I

Analyze and Compare Bags of Cubes

After the class has solved the Jelly Bean Bag Mix-up, write about the investigation by answering these questions.

1 Write about your group's bag of cubes.

a. Which bag did your group have? What strategies did your group use to try to figure this out?

b. Use your group's data to figure out the experimental probability of picking a cube of each color from the bag.

c. What is the theoretical probability of picking a cube of each color from the bag?

2 Compare the five bags of cubes.

a. Rank the five bags from best to worst theoretical probability of picking a red cube. (Best = 1, Worst = 5)

b. Rank the five bags from best to worst theoretical probability of picking a green cube. (Best = 1, Worst = 5)

c. Explain how you figured out how to rank the bags.

d. Fiona took many samples from one of the bags. She got 62 reds, 41 blues, 8 yellows, and 9 greens. Which bag or bags do you think she had? Why?

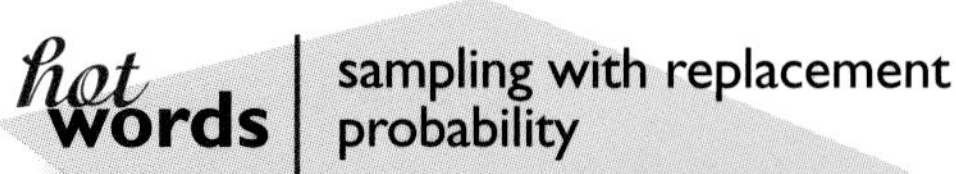

page 47

Class Survey

Applying Skills

In items 1–5, find the range and mode (if any) for each set of data. Be sure to express the range as a difference, not as an interval.

1. 14, 37, 23, 19, 14, 23, 14

2. 127, 127, 117, 127, 140, 133, 140

3. 93, 40, 127, 168, 127, 215, 127

4. 12, 6, 23, 45, 89, 31, 223, 65

5. 1, 7, 44, 90, 6, 89, 212, 100, 78

6. Mr. Sabot's class took a survey in which students were asked how many glasses of water they drink each day. Here are the results:

Glasses of Water Students Drink

X = one student's response

0	1	2	3	4	5	6	7	8	9	10
					X					
				X	X	X				
				X	X	X				
			X	X	X	X	X			
			X	X	X	X	X			
			X	X	X	X	X	X		

Glasses per day

What are the range and mode of the data?

7. Ms. Feiji's class took a survey to find out how many times students had flown in an airplane. Below is the data. Make a frequency graph for the survey and find the mode and range.

- Nine students had never flown.
- Ten students had flown once.
- Six students had flown twice.
- One student had flown five times.

Extending Concepts

8. Ms. Olvidado's class took this survey, but they forgot to label the graphs. Decide which survey question or questions you think each graph most likely represents. Explain your reasoning.

Question 1: How many hours do you sleep on a typical night?

Question 2: How many times do you eat cereal for breakfast in a typical week?

Question 3: In a typical week, how many hours do you watch TV?

Mystery Graph 1

X = one student's response

0	1	2	3	4	5	6	7	8	9	10	11
								X			
								X	X		
								X	X		
							X	X	X		
							X	X	X		
							X	X	X		
							X	X	X		
						X	X	X	X		
						X	X	X	X	X	
						X	X	X	X	X	

Mystery Graph 2

X = one student's response

0	1	2	3	4	5	6	7
				X			
				X			X
X				X			X
X				X	X	X	X
X				X	X	X	X
X		X	X	X	X	X	X
X	X	X	X	X	X	X	X

Writing

9. Answer the letter to Dr. Math.

Dear Dr. Math:

We tried to survey 100 sixth graders to find their preferences for the fall field trip, but somehow we got 105 responses. Not only that, some kids complained that we forgot to ask them. What went wrong? Please give us advice on how to conduct surveys.

Minnie A. Rohrs

Name Exchange

Applying Skills

Find the mean and median of each data set.

1. 10, 36, 60, 30, 50, 20, 40
2. 5, 8, 30, 7, 20, 6, 10
3. 1, 10, 3, 20, 4, 30, 5, 2
4. 18, 22, 21, 10, 60, 20, 15
5. 29, 27, 21, 31, 25, 23
6. 3, 51, 45, 9, 15, 39, 33, 21, 27
7. 1, 4, 7, 10, 19, 16, 13
8. 10, 48, 20, 22, 57, 50

A study group has the following students in it:

Girls: Alena, Calli, Cassidy, Celina, Kompiang, Mnodima, and Tiana

Boys: Dante, Harmony, J. T., Killian, Lorn, Leo, Micah, and Pascal

9. Find the mean and median number of letters in the girls' names.
10. Find the mean and median number of letters in the boys' names.
11. Find the mean and median number of letters in *all* the students' names.
12. Find the mean and median numbers of pretzels in a bag of Knotty Pretzels, based on this graph of the results of counting the number of pretzels in 10 bags.

Knotty Pretzels
X = one bag

```
          X
          X
          X        X
X    X    X    X   X    X
------------------------------
148  149  150  151 152  153
     Number of pretzels
```

Extending Concepts

Professor Raton, a biologist, measured the weights of capybaras (the world's largest rodent) from four regions in Brazil.

Weights of Capybaras

Region	Weights (kg)
A	6, 21, 12, 36, 15, 12, 27, 12
B	18, 36, 36, 27, 21, 48, 36, 33, 21
C	12, 18, 12, 21, 18, 12, 21, 12
D	30, 36, 30, 39, 36, 39, 36

13. Find the mean and median weight of the capybaras in each region.
14. Find the mode and range of each data set. For each set explain what the range tells us that the mode doesn't.

Writing

15. Answer the letter to Dr. Math.

Dear Dr. Math:
When we figured out the mean, median, mode, and range for our survey, some answers were fractions or decimals, even though we started with whole numbers. Why is this? If the numbers in the data set are whole numbers, are *any* of those four answers *sure* to be whole numbers?
Frank Shun and Tessie Mahl

TV Shows

Applying Skills

Jeff conducted a survey rating TV shows on a scale of 1 (“bo-o-o-oring”) to 5 (“Excellent, dude!”). Here are the results:

Show 1

Rating	Number of Students
1	0
2	3
3	7
4	3
5	7

Show 2

Rating	Number of Students
1	2
2	2
3	4
4	7
5	5

1. Draw a frequency graph of the results of each survey.
2. Find the mode(s) for the ratings, if any, of each survey.
3. Find the median rating for each survey.

Lara’s class took a survey asking students to rate four different activities on a scale of 1 (“Yuck!”) to 5 (“Wowee!”).

Activity A
X = one student’s response

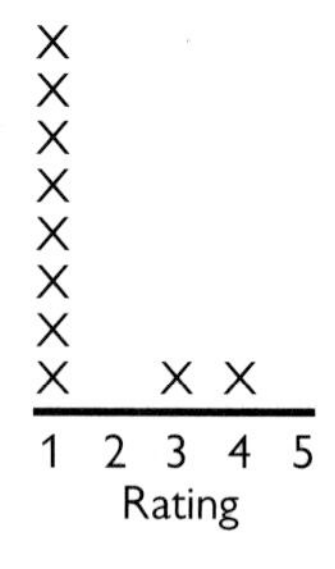

Activity B
X = one student’s response

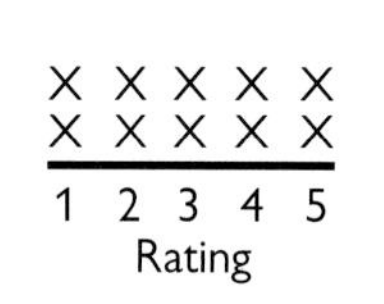

Activity C
X = one student’s response

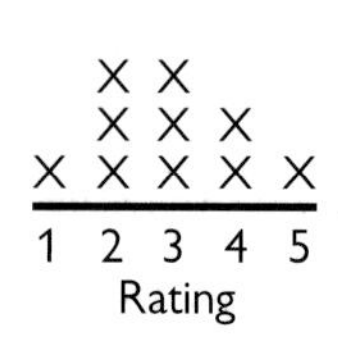

Activity D
X = one student’s response

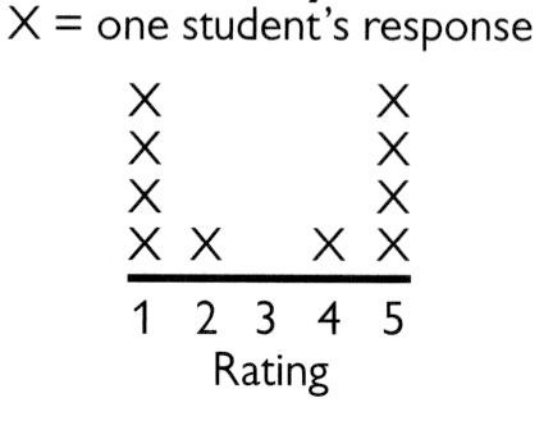

4. Find the median rating for each survey.
5. Find the mean rating for each survey.

Extending Concepts

Statisticians describe graphs of data sets by using four different types of distributions.

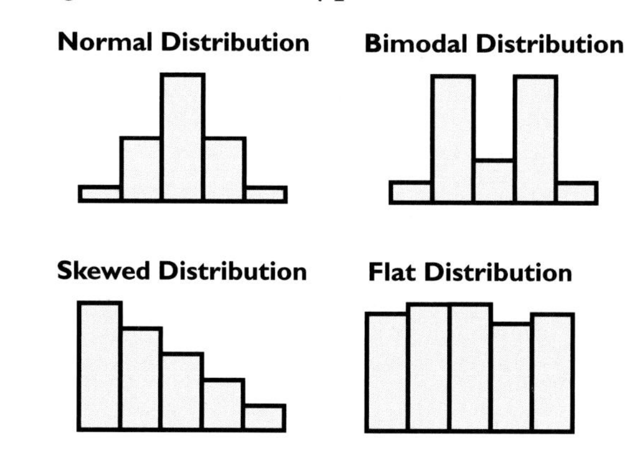

6. Describe the shape of each graph showing the data from Lara’s class.
7. Here are four activities: going to the dentist, listening to rap music, taking piano lessons, and rollerblading. Tell which activity you think goes with each graph for Lara’s class, and why.

Animal Comparisons

Applying Skills

Here is some information about dinosaurs. “MYA” means “Millions of Years Ago.”

Dinosaur	Length (ft)	Height (ft)	Lived (MYA)
Afrovenator	27	7	130
Leaellynasaura	2.5	1	106
Tyrannosaurus	40	18	67
Velociraptor	6	2	75

1. Make a bar graph showing the length of each dinosaur.
2. Make a bar graph showing the height of each dinosaur.
3. Make a bar graph showing how many millions of years ago the dinosaurs lived.

Here are one student’s answers to items **1–3:**

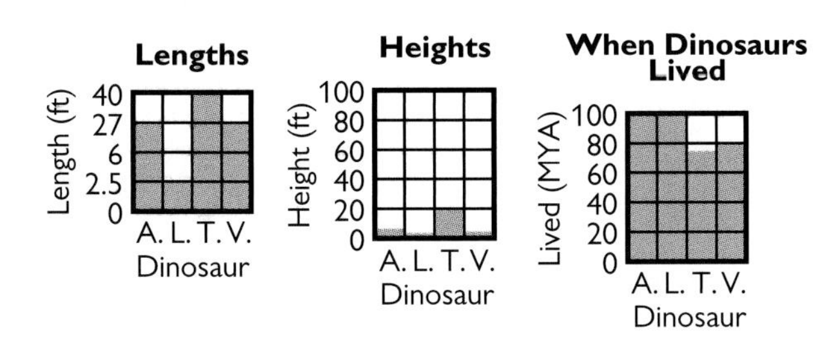

4. What is wrong with the bar graph of the dinosaurs’ lengths?
5. What is wrong with the bar graph of the dinosaurs’ heights?
6. What is wrong with the graph showing how long ago the dinosaurs lived?

Extending Concepts

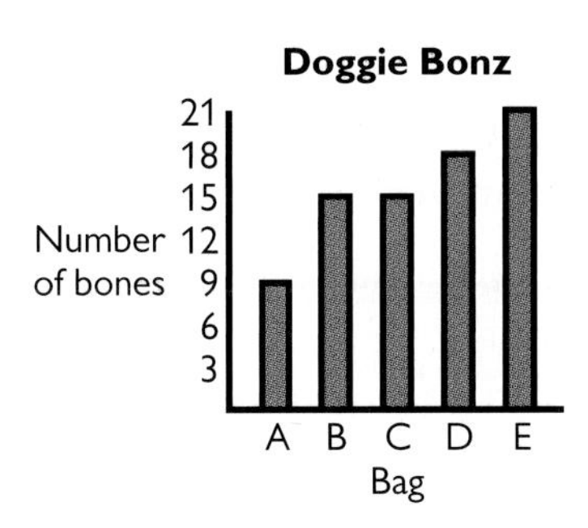

7. Use the graph that shows the number of dog bones in five different bags of Doggie Bonz to make a table of the data.
8. Find the range, mean, and median number of bones in a bag of Doggie Bonz.

Making Connections

Some scientists think that the size of the largest animals on land has been getting smaller over many millions of years. Here are the weights of the largest *known* animals at different periods in history.

Animal	MYA	Estimated Weight (tons)
Titanosaur	80	75
Indricothere	40	30
Mammoth	3	10
Elephant	0	6

9. Draw a bar graph of this information.
10. Does the graph seem to support the conclusion that the size of the largest animals has been getting smaller? What are some reasons why this conclusion might *not* actually be true?

Double Data

Applying Skills

Forty middle school students and 40 adults were asked about their favorite activities. Here are the results of the survey.

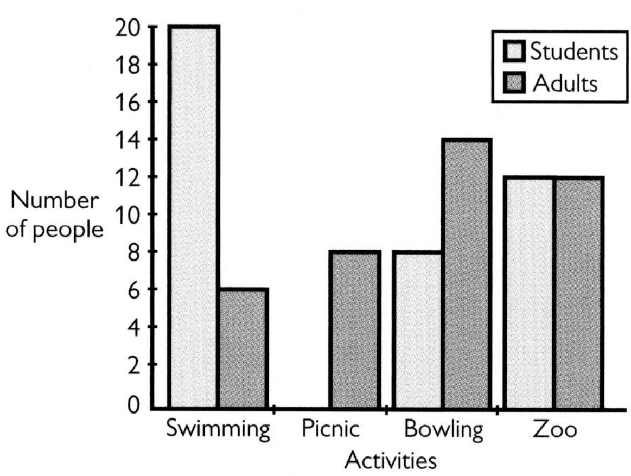

1. Show the survey results in a table (like the one for items 5–7 below).

2. What is the most popular activity for students?

3. What is the most popular activity for adults?

4. Why is there no student bar for Picnic?

Eighty 6-year-olds and eighty 12-year-olds were asked how many hours a day they usually watch TV. Here are the results.

Hours	6-Year-Olds	12-Year-Olds
0	16	4
1	23	9
2	26	15
3	15	38
4	0	14

5. Make a double bar graph of the data.

6. Calculate the mean, median, and mode for the number of hours each group watches TV.

7. Which group on average watches more TV?

Extending Concepts

Signorina Cucina's cooking class rated pies made with 1 cup of sugar, 2 cups of sugar, or 3 cups of sugar. Here are the results.

	Yucky	OK	Yummy
1 cup	2	4	14
2 cups	7	9	4
3 cups	11	6	3

8. Make a triple bar graph of the results. Label the y-axis *Number of Students* and the x-axis *Number of Cups.*

9. Now make another triple bar graph with the survey results. This time, label the y-axis *Number of Students* and the x-axis *Yucky*, *OK*, and *Yummy.*

Writing

10. Tell whether you could make a double bar graph for each set of data. If you *could* make one, tell what the labels on the axes would be. If not, explain why not.

a. Heights of students at the beginning of the year and at the end of the year.

b. Ages of people who came to see the school show.

Across the Ages

Applying Skills

Here are the ratings given to two different bands by 100 students and 100 adults.

Band A

Rating	Number of Students	Number of Adults
Terrible	3	1
Bad	5	3
OK	40	27
Good	43	60
Great	9	9

Band B

Rating	Number of Students	Number of Adults
Terrible	2	26
Bad	10	22
OK	14	20
Good	21	18
Great	53	14

1. Make a double bar graph for the ratings of Band A. Use different colors for students and adults.

2. Make a double bar graph for the ratings of Band B. Use different colors for students and adults.

3. Make a double bar graph for the ratings by students. Use different colors for Band A and Band B.

4. Which band would be best for a party for students?

5. Which band would be best for a party for adults?

6. Which band would be best for a party for students and adults?

Extending Concepts

7. Here are the results of a survey a student did on the number of glasses of milk 35 sixth graders and 35 adults drink in a typical week. Describe what's wrong with the graph and make a correct one.

Glasses per Week	Number of 6th Graders	Number of Adults
0	2	7
1	1	0
2	0	1
3	2	7
4	1	0
5	4	7
6	5	4
7	10	6
8	4	2
9	6	1

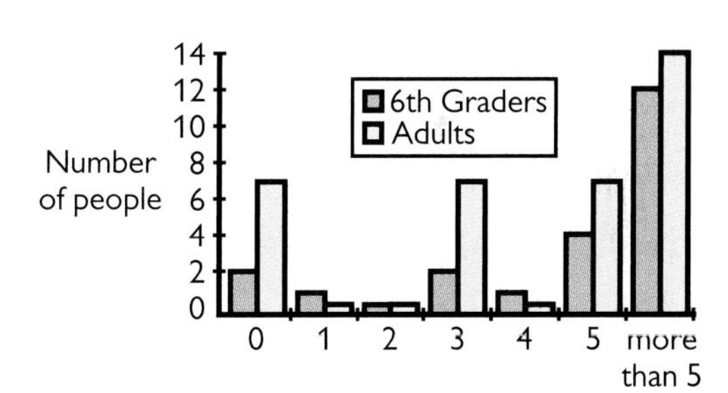

Writing

8. Pat wanted to compare how many seventh graders and kindergartners have pets. She found that of 25 seventh graders, 15 had pets. She didn't know any kindergartners, so she questioned 5 younger brothers and sisters of the seventh graders. Two of them had pets. What do you think about Pat's survey?

Are You Improving?

Applying Skills

Our Data for the Memory Game

	Number of Objects Remembered	
Trial	**Ramir**	**Anna**
1	3	6
2	4	5
3	6	5
4	7	8
5	8	7

The table shows how Ramir and Anna did when they played the Memory Game.

1. Draw a broken-line graph to show each student's progress. Use a different color to represent each student.

Find the following information for Ramir and for Anna.

2. Find the median number of objects remembered correctly.

3. Find each mean.

The graph shows Caltor's progress while playing the Memory Game.

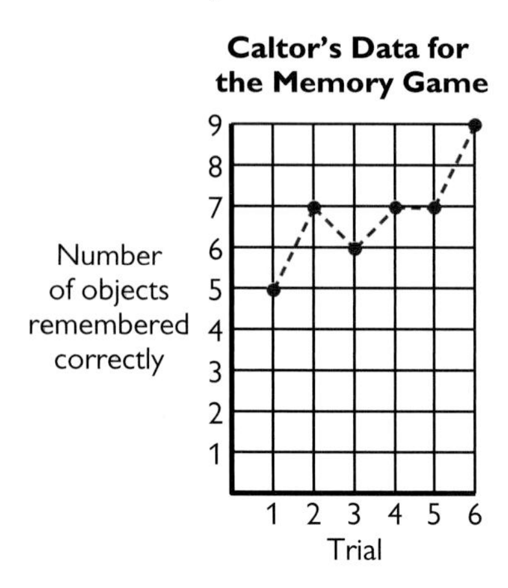

4. How many objects did Caltor remember correctly on Trial 3?

5. How many more objects did Caltor remember correctly on the 6th trial than on the 1st trial?

6. What is the mode?

Extending Concepts

Katia is trying to learn Spanish. Her teacher gave her worksheets with pictures of 20 objects. She has to write the Spanish word for each object. Then she checks to see how many words she got correct. The graph shows her progress.

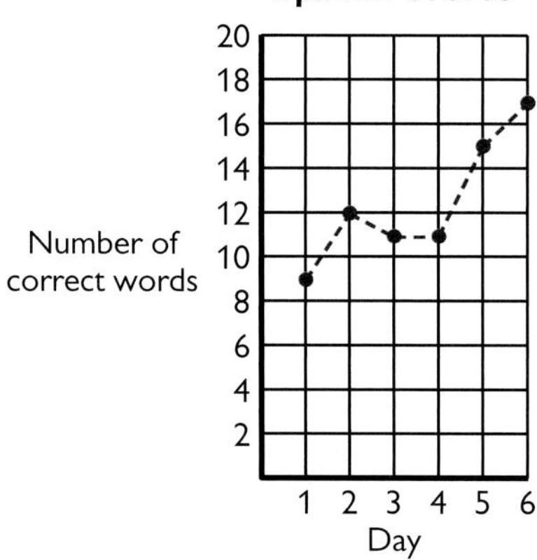

7. Create a table of the data.

8. Find the mean, median, mode, and range for Katia's data.

Writing

9. Give examples of five different types of data for which you might use a broken-line graph. Tell how you would label the x-axis and the y-axis for each graph.

How Close Can You Get?

Applying Skills

The table below shows the results for a student who played the Open Book Game five times.

Dolita's Data for the Open Book Game

Trial	Target	Estimate	Error
1	432	334	
2	112	54	
3	354	407	
4	247	214	
5	458	439	

1. Complete the table by finding the Error for each trial.

2. Make a broken-line graph of Dolita's errors.

3. Find the mean, median, mode (if any), and the range of Dolita's errors.

Every day for 12 days, Tomas runs down the block and times how long it takes.

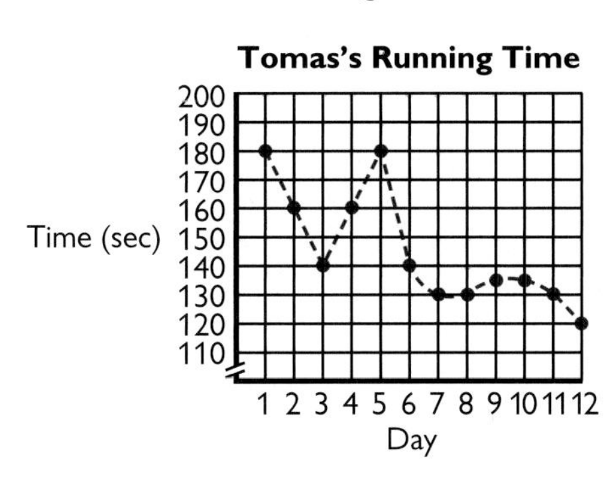

4. Use the graph to figure out the mode, median, and range for Tomas's running times.

5. Use a calculator to figure out Tomas's mean time for running down the block.

6. Make a prediction for how fast you think Tomas would run on Day 15. Explain how you made your prediction.

7. Look at the data for the first 5 days only. On which days did Tomas get faster?

8. On which day(s) did Tomas run the fastest?

Extending Concepts

9. The graphs shown represent four students' progress in the Open Book Game. For each graph, write a description of the student's progress.

10. Which graph shows the least improvement? Explain.

11. Which graph shows the most improvement? Explain.

Writing

12. Dottie noticed that when her scores in the Memory Game improved, her graph kept going up. But when she played the Open Book Game, her graph went down, although she was sure she was improving. Explain to Dottie how to read a broken-line graph.

Stories and Graphs

Applying Skills

A.

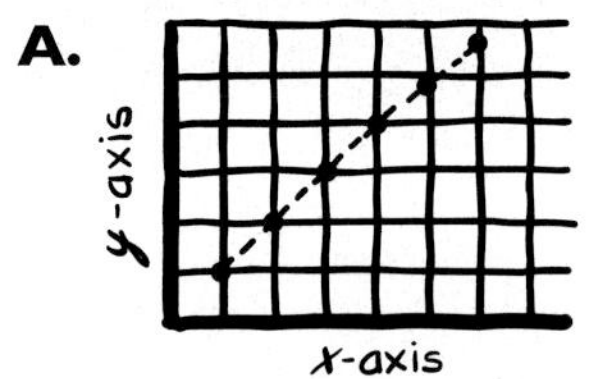

B.

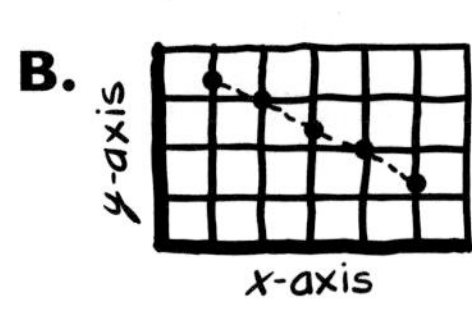

C.

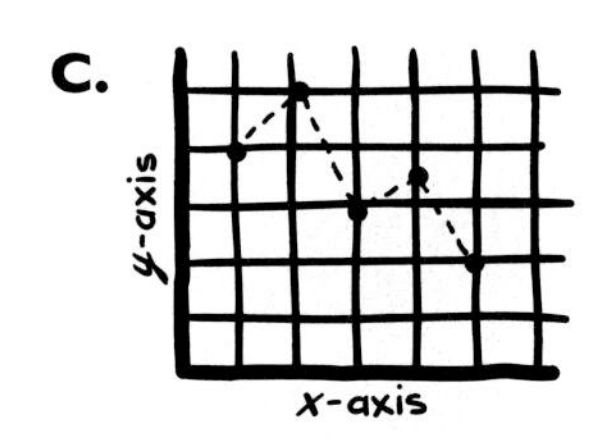

D.

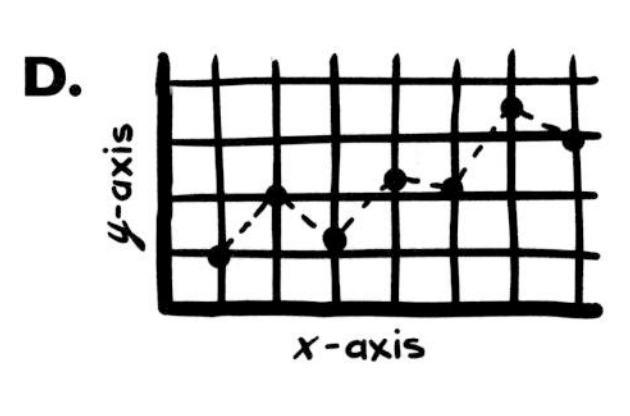

Match each description with the appropriate graph. Then tell how you would label the *x*-axis (across) and the *y*-axis (up and down).

1. "I've made steady improvement in the Memory Game."
2. "I've been running faster every day."
3. "My errors for the Open Book Game have been going up and down. Overall, I've gotten better."
4. "My swimming speed has been going up and down. Overall, I don't seem to be improving!"
5. "I've been timing how long I can stand on my head. I've had good days and bad days, but mostly I've increased my time."

Extending Concepts

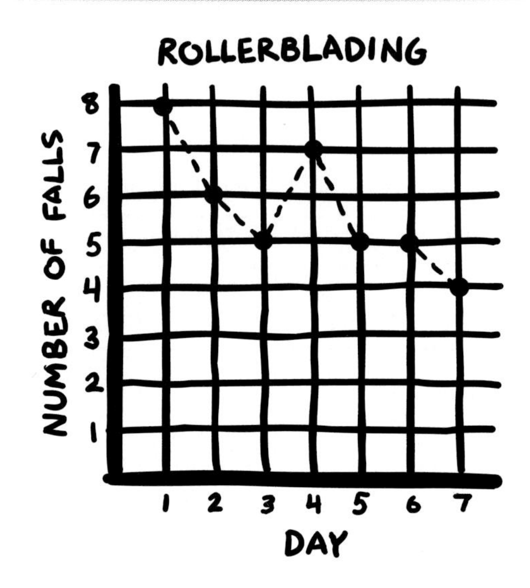

6. Make a table for the data represented on the graph "Rollerblading."
7. Calculate the mean, median, mode (if any), and range for number of falls.
8. Did this student improve at rollerblading? Write a sentence to describe his progress.

Writing

9. Answer the letter to Dr. Math.

Dear Dr. Math,
I'm confused. I don't know how you can look at a broken-line graph that has no numbers and figure out whether it shows that a student is or isn't improving.
Reada Graph

What Are the Chances?

Applying Skills

Each student picked a cube from a bag twenty times. After each turn, the cube was returned to the bag. Results for each student were recorded in the table.

Data From Our Experiment

Students	Number of Greens	Number of Yellows
Anna	15	5
Bina	18	2
Carole	12	8
Dan	16	4
Elijah	14	6

Use the data on green and yellow cubes to find the following values for each color.

1. mode **2.** range

3. mean **4.** median

Use fractions to describe each student's experimental probability of getting a green cube.

Example: Anna: $\frac{15}{20}$

5. Bina **6.** Carole **7.** Dan

8. Elijah

Use fractions to describe each student's experimental probability of getting a yellow cube.

9. Anna **10.** Bina **11.** Carole

12. Dan **13.** Elijah

14. Combine the data for the whole group. What is the whole group's experimental probability for picking a **green cube**?

Extending Skills

Here is a list of bags that the students might have used to collect the data shown in the table. For each bag, decide whether it is likely, unlikely, or impossible that students used that bag. Explain your thinking.

15. 28 green, 12 yellow

16. 10 green, 30 yellow

17. 8 green, 2 yellow

18. 16 green, 4 blue

19. 15 green, 15 yellow

20. If students used a bag with 100 cubes in it, how many green and yellow cubes do you think it contained? Explain your thinking.

Making Connections

21. Frequently on TV the weather reporter gives the chance of rain as a percentage. You might hear, "There's a 70% chance of rain for tomorrow afternoon, and the chances increase to 90% by tomorrow night." What does this mean? Why do you think this kind of language is used to talk about weather? In what other situations do people talk about the chances of something happening?

Changing the Chances

Applying Skills

Bag	Blue	Red	Total Number of Cubes	Theoretical Probability of Picking Blue	Theoretical Probability of Picking Red
A	9	1	10	$\frac{9}{10}$	$\frac{1}{10}$
B	7	13			
C	16	4			
D	15	15			
E	22	8			
F	30	10			

1. Copy the table and fill in the missing information. Use the first row as an example.

2. Which bag gives you the highest probability of getting a blue cube?

3. Which bag gives you the highest probability of getting a red cube?

4. Which bag gives you the same chance of picking a blue or a red cube?

5. Yasmine has a bag with 60 cubes that gives the same probability of picking a blue cube as Bag C. How many blue cubes are in her bag?

6. How many red cubes are in Yasmine's bag?

7. Rank the bags from the best chance of getting a blue cube to the worst chance of getting a blue cube.

Extending Concepts

Students did experiments with some of the bags shown in the table. The results of these experiments are given below. For each of the results, find the indicated experimental probability. Which bag or bags do you think it is most likely that the students used? Why?

8. In 100 turns, we got 20 reds.

9. We got 44 blues and 46 reds.

10. In 100 turns, we got 75 blues.

11. In 5 turns, we got 0 reds.

Writing

12. Suppose Sandy's bag has 2 purple cubes out of a total of 3 cubes and Tom's bag has 8 purple cubes out of 20 cubes. Explain how to figure out which bag gives you the best chance of picking a purple cube if you pick without looking.

Which Bag Is Which?

Applying Skills

This bar graph shows the number of cubes of different colors that are in a bag of 20 cubes.

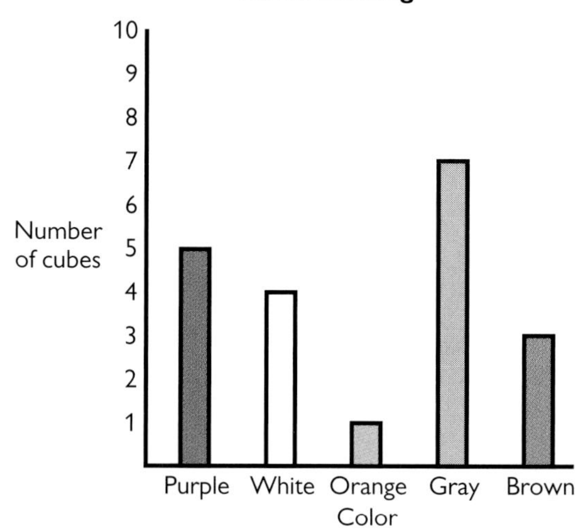

Use the graph to figure out the theoretical probability of picking each color. Be sure to write the probability as a fraction.

1. a purple cube

2. a white cube

3. an orange cube

4. a grey cube

5. a brown cube

Students each took 10 samples from the bag and recorded their data in the table shown.

Data From Our Experiment

Student	Purple Cubes	White Cubes	Orange Cubes	Gray Cubes	Brown Cubes
Miaha	2	2	0	4	2
Alec	3	1	1	5	0
Dwayne	3	2	1	3	1
SooKim	2	3	0	3	2

6. Combine the data for all the students to figure out the group's **experimental probability** of picking each color. Write the probability as a fraction.

7. For each color, find the mean number of times it was picked.

Extending Skills

8. A box of Yummy Chewy Candy has 30 pieces of candy. The pieces of candy are blue, green, red, and pink. The theoretical probability of picking a blue piece is $\frac{1}{3}$, a green piece is $\frac{1}{6}$, and a red piece is $\frac{1}{5}$. How many pieces of pink candy are in the box? Explain.

Writing

9. Answer the letter to Dr. Math.

Dear Dr. Math,

I was looking at the results of the Jelly Bean Supreme Investigation and I'm confused. The theoretical probability of picking a blueberry from Bag A is $\frac{7}{12}$. My group picked 24 times from Bag A and got 16 blueberries. Is that more or less blueberries than you would expect? Why didn't our results match the theoretical probability exactly?

Beanie

The McGraw-Hill Companies

This unit of MathScape: Seeing and Thinking Mathematically was developed by the Seeing and Thinking Mathematically project (STM), based at Education Development Center, Inc. (EDC), a non-profit educational research and development organization in Newton, MA. The STM project was supported, in part, by the National Science Foundation Grant No. 9054677. Opinions expressed are those of the authors and not necessarily those of the Foundation.

CREDITS: Unless otherwise indicated below, all photography by Chris Conroy and Donald B. Johnson.

3 (tr)courtesy KTVU-TV (Bill Martin); **16 17** Image Club Graphics; **17** (tr)courtesy KTVU-TV (Bill Martin).

Copyright ©2005 by The McGraw-Hill Companies, Inc. All rights reserved. Printed in the United States of America. Except as permitted under the United States Copyright Act, no part of this publication may be reproduced or distributed in any form or by any means, or stored in a database or retrieval system, without prior permission of the publisher.

Send all inquiries to:
Glencoe/McGraw-Hill
8787 Orion Place
Columbus, OH 43240-4027

ISBN: 0-07-866792-5

1 2 3 4 5 6 7 8 9 10 058 06 05 04 03